I marched in a band last year. We played songs while we marched. People always cheered and clapped.

I played the drums in the marching band. I banged my drum with two sticks. I stepped and stomped to the beat.

My friend Matt played the sax. He could play so many songs! He tooted his sax while we marched. I wished I could play the sax too.

We worked hard to be good. Some days we just played the music. Other days we just marched. Then we did both at the same time.

Once we were going to play at a show. It was going to be fun for us. There would be lots of other bands.

We all rode the bus to the show. I was sitting next to my friend Meg. Meg yelled out, "I can't find my trumpet!"

We all looked for her trumpet. But it was useless. Meg had messed up and left it at home. She felt helpless.

Before the show, we met other band members. A boy from another band loaned Meg his trumpet. Meg blasted the trumpet and I banged the drums. Our marching band sounded great!